SOTHIS PRESS

Sothis Press aims to completely reimagine the public domain book with an emphasis on universal access to captivatingly designed print editions. We believe that technology will enable every single book ever published to be perpetually accessible in beautiful print form. Sothis is beauty in ubiquity.

www.sothis.press
1 Waterhouse Square, London, EC1N 2ST

Electroculture

Lyman James Briggs

SOTHIS PRESS

ISBN 978-1-83548-396-1

Electroculture is in the
public domain.

This edition is published by
Sothis Press © 2023. The company
is registered in England and Wales.
The Company Number is 14900389.

UNITED STATES DEPARTMENT OF AGRICULTURE

DEPARTMENT BULLETIN NO. 1379

Washington, D. C. January, 1926

ELECTROCULTURE

By Lyman J. Briggs,[1] A. B. Campbell, R. H. Heald, and L. H. Flint, *Office of Biophysical Investigations, Bureau of Plant Industry*

CONTENTS

The term "electroculture" as used in this bulletin refers to practices designed to increase the growth and yield of crops through electrical treatment, such as the maintenance of an electric charge on a network over the plants or an electric current through the soil in which the plants are growing.

During the past 75 years many experiments in electroculture have been carried out with varying degrees of refinement. Some of these experiments indicate that the yield of crops can be materially increased by electrical treatment. Others, conducted along similar lines, fail to show any marked response to the treatment. In this latter class are included the experiments conducted by the Office of Biophysical Investigations of the Bureau of Plant Industry, which are reported in the following pages. This report is followed by a brief account of other investigations in this field. Investigations relating to the cultivation of plants under electric lights are not included in the review of the literature of electroculture, the response of the plants under such conditions being due primarily to the heat and light into which the electrical energy has been transformed.

NORMAL ELECTRICAL STATE OF THE ATMOSPHERE

Since the effect of using a charged network over growing plants is to change the electrical state of the atmosphere surrounding the plants it seems desirable to discuss briefly the normal electrical conditions in the atmosphere and the changes produced by the charged network. An examination of the electrical conditions in the atmosphere over an open field on a clear day shows that there is a force tending to move a positively charged body downward; in other words, the electrical field of force is identical with that which would exist if the earth were charged negatively.

[1] Physicist, Bureau of Standards, since 1920.

62149°—26†——1

On fine days, the potential gradient in the atmosphere is almost invariably positive in sign (that is, a positive charge tends to move downward), and the magnitude of the vertical gradient is of the order of 100 volts per meter, though it is continually varying. When thunderstorms are in the neighborhood, the potential gradient may be either positive or negative and changes sign frequently. The magnitude of the potential gradient also undergoes wide fluctuations, during stormy weather frequently attaining values of 10,000 volts per meter, 100 times the normal gradient.

A further examination of the lower atmosphere shows that charged particles or ions are always present. Both positive and negative ions are found, the positive ions generally being somewhat more numerous. They consist of groups of molecules loosely bound together and carrying a charge. Frequently these small ions attach themselves to dust particles, thus becoming large ions, which move much less rapidly than the small ions.

When the potential gradient is positive, the negative ions move upward and the positive ions downward to the ground, thus constituting an electric current flowing from air to earth. This current is due almost entirely to the small or free ions, the mobility of the large ions being so low that their influence on the conductivity of the air can be disregarded. The magnitude of the current from the air to a unit area on the earth's surface is extremely small, being only 2×10^{-12} amperes per square meter or 5×10^{-8} amperes per acre. The strength of the current is proportional to the potential gradient, to the number of ions per unit volume, and to their mobility. The average number of free ions is of the order of 1,000 per cubic centimeter, the positive ions constituting somewhat more than one-half the total number. Their mobility is such that they migrate with a velocity of about 1 centimeter per second when subjected to a potential gradient of 100 volts per meter.

Although the air-earth current per unit area is extremely small, it is sufficient when applied to the whole of the earth's surface to reduce the negative charge of the earth to one-half its initial value in about 10 minutes. The explanation of the maintenance of the negative charge of the earth under such extraordinary conditions is one of the outstanding problems in atmospheric electricity (*12, 47*).[2]

ELECTRICAL FIELD EMPLOYED IN ELECTROCULTURAL EXPERIMENTS

In most of the field experiments conducted at the Arlington Experiment Farm, the standard height of the network was 5 meters, and the potential of the network was approximately 50,000 volts. The average potential gradient under the network was therefore of the order of 10,000 volts per meter, or about 100 times the normal gradient in fine weather. This would produce an air-earth current about 100 times the normal current as long as the ion content of the air remained normal. However, a marked ionization occurred at the network, so that the number of positive ions per unit volume under the network was much higher than normal. This was shown by means of measurements made when the network was charged and a gentle breeze blowing. On the windward side of the network the conditions were normal, but on the leeward side a decided increase

[2] The serial numbers (*italic*) in parentheses refer to "Literature cited," at the end of this bulletin.

was observed in the ion content of the air which drifted from under the network. This effect could be traced to a distance of several hundred feet from the network.

The principal change in the environment of plants grown under a charged network appears then to consist in a marked increase in the strength of the air-earth current which flows through the plants to the ground.

If the drifting charge from the experimental plat should pass over the control plat, it would increase the air-earth current to the control plat to some extent, owing to the increase in the number of ions per unit volume. But even under such conditions the current flowing into the control plat would necessarily be small in comparison with that flowing into the experimental plat, since both the ion content and the potential gradient are much higher under the network and the current is proportional to the product of these factors.

ELECTROCULTURAL EXPERIMENTS WITH MISCELLANEOUS CROPS

Experiments in 1907.—Electrocultural experiments were first undertaken by the department [3] in 1907, using vegetables for the most part as test crops. The test plat, which was 138 by 106 feet, was divided into three sections 44 by 106 feet, the center section being used as the experimental area and the two outside sections as controls. The crops were planted in continuous rows across the three sections, so that the center third of each row was under treatment.

A Wagner mica-plate electrostatic machine was used as a high potential source. It was inclosed in a tight case, permitting the use of drying agents to keep the machine in the best condition for operation. The positive pole was connected to an open wire network strung on glass insulators, and the negative pole was grounded. The network covered the experimental plat and was placed high enough to permit the use of a horse cultivator. The applied potential varied somewhat with weather conditions, but usually exceeded 50,000 volts. The network was charged throughout the night, from late afternoon until early morning. The plants were subjected to the electrical treatment 656 hours in all, extending from June 20 to September 16. The yields are shown in Table 1.

TABLE 1.—*Yields following electrocultural treatment of miscellaneous crops under test at Arlington Experiment Farm in 1907*

Crop	Yields per plat (pounds)				Ratio of treated to average of controls
	Experimental plat	Control		Average of controls	
		Plat A	Plat C		
Tomatoes	128.25	119.75	138.75	129.25	0.996
Cowpeas	20.60	19.14	24.25	21.70	.95
Cowpea vines	30.9	40.14	43.0	41.57	.742
Potatoes	44.0	52.0	39.0	45.5	.968
Turnips	55.0	45.0	64.0	54.5	1.01
Beets	24.0	25.0	25.0	25.0	.96
Carrots	15.0	18.0	17.0	17.5	.856
Cabbage	106.05	106.0	80.0	93.0	1.14
Buckwheat	20.0	25.0	23.0	24.0	.835
Beans	8.5	10.0	11.5	10.75	.790

[3] These experiments were conducted on the Arlington Experiment Farm by the Office of Biophysical Investigations and the Office of Crop Physiology and Breeding Investigations, the field work being handled largely by E. W. Hudson and W. Seifriz.

The lack of uniformity in the yields of the control plats A and C in the 1907 experiments (Table 1) is such that no great dependence can be placed in these results. It is significant, however, that in only one of the 10 trials recorded did the treated plat show any evidence of a substantial increase in yield when compared with the mean of the control plats.

Experiments in 1908.—In the 1908 trials the wires were run directly over the treated rows and kept at a height of 6 to 18 inches above the plants by means of adjustable brackets on which the insulators were mounted. The control rows ran parallel to the treated ones at a distance of 6½ feet and were separated from them by intermediate guard rows.

In one part of the plat the wires over the plants were charged positively to about 50,000 volts from 4 p. m. to 7 a. m. each day, 955 hours in all. In the other part of the plat the wires were charged and discharged rapidly by connecting them to one terminal of the secondary of an induction coil, the other terminal being grounded. In this case the potential rose to about 20,000 volts and then discharged suddenly through a small spark gap between the wires and the ground.

The treatment first described is similar to that employed by Lemström and believed by him to result in increased yields. In these experiments, however, neither treatment gave any evidence of increased growth. The detailed yields consequently are not of special interest.

ELECTROCULTURAL FIELD EXPERIMENTS WITH GRAINS

In selecting a location for the electrocultural field experiments near Washington, three conditions were sought: (1) A uniform soil, (2) available electric power, and (3) accessibility from the laboratory in Washington, since the equipment had to be visited daily during the experimental season. Soil uniformity is particularly difficult to find in the environs of Washington, and the Arlington Experiment Farm forms no exception in this respect. It seemed to be the best available location, however, and portions of sections A, B, and E were made available for the experiments, which were carried on from 1911 to 1918. Sections A and B proved very disappointing with regard to their uniformity, and the most reliable results were obtained in section E. These experiments will be first described.

The Lodge-Newman apparatus used in the experiments from 1912 to 1915, inclusive, was designed in England primarily for electrocultural work and consists essentially of a 110-volt induction coil, operated by a mercury interrupter, and a rectifier. Five Lodge valves [4] designed to rectify the high-tension alternating current were placed in series with the network, thus allowing only the positive impulses from the secondary of the coil to reach the network (*33*). The negative pole was grounded. Two balls 25 millimeters in diameter, one of which was grounded and the other connected to the network, were used to determine the potential, assuming a breakdown gradient of 3,000 volts per millimeter.

Systematic measurements of the current from the network were not made, but the current could be determined approximately from the potential of the network and the known power characteristics of

[4] For a description of the valves, see Lodge, O. (*34*).

the machine used. The current from the network over the experimental plat in section E was of the order of 0.1 to 1 milliampere per acre, depending on the voltage and network used. This is of the order of 10,000 to 100,000 times the intensity of the normal air-earth current.

EXPERIMENTS IN SECTION E

It has been shown by Jørgensen and Priestley (*26*) that the ionization from the highly charged network is by no means limited to the area beneath the network, but may be carried by the wind to a considerable distance, depending on the weather conditions. It was consequently deemed advisable to separate the treated and control plats so far as practicable. Accordingly, two plats of half an acre each (132 by 165 feet) were selected in section E which were separated by a distance of 350 feet, one plat being directly north of the other.

FIG. 1.—General view of the experimental field at Arlington Experiment Farm, showing the system of double insulators used in suspending the wire network from poles and the power lines leading to the motor in the apparatus house (foreground). Poles supporting the grounded network along the side of the control plat may be seen in the distance. (Photographed May 8, 1918.)

The rye which was growing on the plats of section E when they were selected in 1913 was cut and weighed. The results show that the productiveness of the two plats was about the same, being as follows: Yield of south plat, 2,438 pounds; of north plat, 2,499 pounds; ratio of south plat to north plat 0.98.

Experiments in 1914.—A network 16 feet high was erected over the south plat, having cross wires at intervals of 15 feet. (Fig. 1.) Winter wheat was sown on both plats the following October, and the treatment was given by means of the Lodge-Newman apparatus, which furnished a positive charge to the network at a potential ranging from 30,000 to 60,000 volts. The treatment was given in the fall and spring from 3 to 7 p. m., a total of 336 hours. The grain was harvested in June, 1914, giving yields which were substantially the same for both plats, as shown in Table 2.

TABLE 2.—*Yields of winter wheat on plats following electrocultural treatment (positive charge), section E, Arlington Experiment Farm, in 1914*

Plat	Yields (pounds)		Ratio of treated to control	
	Shock	Grain	Shock	Grain
Treated	2, 332	644. 8	1. 02	0. 97
Control	2, 281	656. 5		

Experiments in 1915.—Wheat was again sown in the autumn of 1914. The fall treatment was omitted, owing to bad weather. In 1915 the network was charged positively by the Lodge-Newman apparatus twice a day from 4 to 7 a. m. and from 5 to 8.30 p. m., a total of 345 hours. The distance between the cross wires of the network this year was 6 feet. The plats were divided at harvest into east and west halves. The yields are shown in Table 3.

In both plats two bad spots developed on the western halves, in which the grain was much poorer than the average.

TABLE 3.—*Yields of winter wheat on plats following electrocultural treatment (positive charge), section E, Arlington Experiment Farm, in 1915*

Plat	Yields (pounds)		Ratio of treated to control	
	Shock	Grain	Shock	Grain
Eastern half:				
Treated	832	321. 5	1. 01	0. 92
Control	822	350		
Western half:				
Treated	716	303	1. 32	1. 19
Control	540	254. 5		
Total:				
Treated	1, 548	624. 5	1. 14	1. 03
Control	1, 362	604. 5		

Experiments in 1916.—In the fall of 1915 winter wheat was again sown, as it was desired to get a test with the network charged negatively, about 45,000 volts, instead of positively as heretofore. A powerful static machine was used to supply the current, and it was run from 4 p. m. to 8 a. m. daily (totaling 800 hours) during the spring, the fall treatment being omitted.

The plats were divided into eastern and western halves at the time of harvest and again showed considerable variation. The yields are given in Table 4.

TABLE 4.—*Yields of winter wheat on plats following electrocultural (negative) treatment, section E, Arlington Experiment Farm, in 1916*

Plat	Yields (pounds)		Ratio of treated to control	
	Shock	Grain	Shock	Grain
Eastern half:				
Treated	1, 324	347. 5	0. 98	0. 85
Control	1, 352	411. 0		
Western half:				
Treated	1, 204	324. 5	1. 10	. 95
Control	1, 092	343. 0		
Total:				
Treated	2, 528	672. 0	1. 03	. 89
Control	2, 444	754. 0		

Experiments in 1917.—Wheat was again sown in section E in October, 1916, and allowed to mature the following summer without treatment, as an additional check on the soil conditions. At time of harvest in 1917 the plats were again cut into eastern and western halves, the south plat being the one which had received the electrical treatment in previous years. The yields are shown in Table 5.

Comparison with the rye yields of 1913 shows that the south (treated) plat apparently gained slightly in its relative productivity during the five years, but the change is well within the errors of field trials.

TABLE 5.—*Yields of winter wheat on plats without electrocultural treatments, section E, Arlington Experiment Farm, in 1917*

Plat	Yields (pounds)		Ratio of south to north plats	
	Shock	Grain	Shock	Grain
Eastern half:				
South plat	1,625.0	580.5	1.00	0.92
North plat	1,625.0	631.0		
Western half:				
South plat	1,562.5	557.0	1.08	.98
North plat	1,439.5	567.5		
Total:				
South plat	3,190.5	1,137.5	1.04	.95
North plat	3,064.5	1,198.5		

Experiments in 1918.—In the fall of 1917 winter wheat (Currell) was sown on the plats in section E, and in the spring a ½-inch mesh galvanized-iron screen 132 feet long by 15 feet high was erected 20 feet south of the check plat. It was thought that the grounded screen might protect the north plat from the drifting charge, but later measurements show that it is of doubtful value.

The static machine was again used, with the positive pole connected to the network. The number of cross wires was increased to one every 3 feet. This increased the current and reduced the potential of the network to about 30,000 volts.

Although the winter was exceptionally cold the stand in the spring was excellent. Treatment was started April 15 and continued for 46 days from 4 p. m. to 8 a. m. each day, a total of 736 hours.

At harvest the eastern and western halves of each plat were kept separate and weighed. The yields are shown in Table 6.

TABLE 6.—*Yields of winter wheat on plats following electrocultural treatment* (positive charge), *section E, Arlington Experiment Farm, in 1918*

Plat	Yields (pounds)		Ratio of treated to control	
	Shock	Grain	Shock	Grain
Eastern half:				
Treated	1,531	569	1.16	1.10
Control	1,332	518		
Western half:				
Treated	1,289	481	.99	.95
Control	1,307	507		
Total:				
Treated	2,820	1,050	1.07	1.02
Control	2,639	1,025		

A general view of the experimental field as it appeared on May 8, 1918, is shown in Figure 1.

After the 1918 crop was harvested, measurements of the charge carried by the wind were undertaken. A flame collector was used, which was connected to the gold leaf of an electroscope, the case being grounded. A full-scale deflection of 25 divisions represented a potential of about 1,000 volts. In all the measurements the collector was held at a height of 1 meter above the ground.

A light south wind was blowing the day the measurements were made. With no charge on the network, a very slight deflection of the gold leaf could be noticed. With the network charged, however, the full-scale deflection occurred very rapidly at any point under and within 20 feet outside the network on all sides, even to the south, the direction from which the wind was coming. At 50 feet south, only about 1 division deflection was obtained. North from the network the deflection to full scale was slower and more irregular the greater the distance from the network, and when only 2 feet south of the screen along the south side of the north plat the maximum deflection obtainable was about 20 divisions. Just north of the grounded screen the maximum deflection obtained was about 9 divisions. As the collector was moved farther north from the screen and into the control plat, the deflection again increased, until at the center of the control plat it was off the scale again. The grounded screen along the south side of the control plat thus afforded little protection from the drifting charge. At a point 1,000 feet from the network, the last point observed, a full-scale deflection was obtained. At all points beyond 100 feet from the network over the south plat the deflection was very irregular and unsteady.

The Weather Bureau records show that during the 46 days of treatment in 1918 the wind was due south only 3 days. Owing to the distance of 350 feet between the treated and control plats, the wind would have to be nearly due south to carry any appreciable charge over the control plat.

SUMMARY OF EXPERIMENTS IN SECTION E

The relative yields of the south (treated) and north plats in section E are summarized in Table 7.

TABLE 7.—*Summary of yields of rye and winter wheat on the south (treated) and north (untreated) plats, section E, Arlington Experiment Farm, in six stated years*

Year	Crop	Treatment of south plat	Ratio of yields of south to north plats		Year	Crop	Treatment of south plat	Ratio of yields of south to north plats	
			Total	Grain				Total	Grain
1913	Rye	None	0.98		1916	Wheat	Negative	1.03	0.89
1914	Wheat	Positive	1.02	0.97	1917	do	None	1.04	.95
1915	do	do	1.14	1.03	1918	do	Positive	1.07	1.02

It is evident from the summary that the electrical treatment did not produce any sensible increase in yield. An examination of the detailed results for 1915 shows that the somewhat higher ratios obtained during this unfavorable year are due to a marked decrease in yield in half of the control plat. Aside from this, there appears to be a gradual increase in the total yield of the south plat relative to the north one, irrespective of whether a positive charge, a negative charge, or no charge at all was used. It is of interest to note that the grain ratios with a positive charge on the network are all slightly higher than the ratio in 1917, when no treatment was given; with the negative charge the reverse is true. This seems consistent, for if increasing the positive gradient of the electrostatic field tends to stimulate growth, then to reverse the sign of the field may perhaps tend to inhibit growth. Opposed to this speculation is the fact that the negative field apparently had no effect on the ratio of the total yields of the two plats. In brief, while there is some evidence of a slight increase in grain yield when wheat is grown under a network which is positively charged to a high potential, the observed effect is so small that it is well within the experimental errors of field trials.

EXPERIMENTS IN SECTION B

Experiments in 1911.—The first electrocultural field experiments at Arlington Experiment Farm were made in 1911 with grains in section B, employing a plat which had been seeded in strips to wheat the previous fall. In the spring of 1911 a network of small wire was installed over the eastern half of the plat, covering half of each variety. The network was 7 feet high with wires at intervals of 3 feet, connected to the positive pole of a static machine operating at a potential of about 40 to 50 kilovolts. The machine was in operation six days a week from 3 p. m. to 7 a. m. except during rainy weather from early spring to harvest.

Table 8 shows the relative yields of the treated and control halves.

TABLE 8.—*Yields of winter wheat on plats following electrocultural treatment (positive charge), section B, Arlington Experiment Farm, in 1911*

Variety	Yields per acre (pounds)				Ratio of grain, treated to control
	Treated half		Control half		
	Grain	Straw	Grain	Straw	
G. I. 1942	820	1,740	780	1,360	1.05
Fultz	1,320	1,920	1,450	2,070	.91
G. I. 1974	1,240	2,520	1,300	----------	.95

Experiments in 1912.—In the fall of 1911 one variety of wheat, Currell (*Currell's Prolific*), was sown on section B, and the network was again erected at the height of 7 feet with cross wires 3 feet apart, as before. The treated and control plats each had an area of three-fourths of an acre. This year the network was charged with a Snook-Roentgen set, which consisted of an inverted rotary converter supplying a 160-volt current to a 1-kilowatt 100,000-volt transformer. A mechanical rectifier was used on the high-tension side to obtain a positive charge on the network, the other terminal of the transformer being grounded. Even with this set it was not possible to charge the network much above 50,000 volts. The treatment was given daily from 3 to 7 p. m., except Sundays and during bad weather.

At harvest the weights shown in Table 9 were recorded.

62149°—26——2

TABLE 9.—*Yields of winter wheat on plats following electrocultural treatment* (positive charge), *section* B, *Arlington Experiment Farm, in 1912*

Plat	Yields (pounds)			Ratio of treated to control		
	Shock	Grain	Straw	Shock	Grain	Straw
Treated	3, 465	1, 154	2. 311	1. 05	1. 04	1. 06
Control	3, 300	1, 114	2, 186			

Experiments in 1913.—In the fall of 1912 the same plat in section B was again sown to wheat. The 7-foot network of the previous year was replaced by a permanent one 16 feet high, with cross wires 10 yards apart. The new network was erected over the northern half of the plat instead of the eastern half as in preceding years. The network was charged positively with the Lodge-Newman apparatus, and the treatment was given daily from 4 p. m. to 8 a. m.

The treated and control portions each had an area of three-fourths of an acre. At harvest the weights shown in Table 10 were recorded.

After the wheat was cut, cowpeas were sown on the B plat on July 29, 1913.

The static machine was connected to the network (16 feet high), giving about 40 to 50 kilovolts. The machine (positive charge) was run four hours a day from 3 to 7 p. m. for 32 days. On account of the lateness of the season, the cowpeas were cut for hay. After being stacked and cured, the crop was weighed in the field by means of a tripod and spring balance, showing the following yields: Treated portion, 1,807 pounds; control portion, 1,847 pounds; ratio of treated to control, 0.98.

TABLE 10.—*Yields of winter wheat on plats following electrocultural treatment* (positive charge), *section* B, *Arlington Experiment Farm, in 1913*

Plat	Yields (pounds)		Ratio of treated to control	
	Shock	Grain	Shock	Grain
Treated	3, 254	808	1. 04	1. 03
Control	3, 139	782		

Experiments in 1914.—Corn was planted in the B plat on May 24, 1914, and the network (16 feet high) was connected directly to one wire of a 6,600-volt 3-phase 25-cycle alternating-current power line running past the farm. The voltage was on continuously day and night for 110 days, when the corn was cut and the total weights recorded in the field. It was then shocked and given time to dry. Husking was done in the field on October 9, 1914, and the grain and fodder brought to a platform balance in the barn and weighed. The superintendent of the farm expressed the opinion that the treated plat had had some advantage over the check plat as regards soil-moisture conditions. The yields shown in Table 11 were recorded.

TABLE 11.—*Yields of corn on plats following electrocultural treatment (alternating charge), section B, Arlington Experiment Farm, in 1914*

Plat	Yields (pounds)			Ratio of treated to control		
	Green shocks	Dry shocks	Grain (on cob)	Green shocks	Dry shocks	Grain (on cob)
Treated	16,031.5	4,060	2,892	1.16	1.03	1.28
Control	13,775.5	3,952	2,260			

Experiments in 1915.—The corn was followed by rye which was sown in section B on October 22, 1914. The 6,600-volt treatment alternating charge was started November 5 and maintained continuously till June 24, 1915. This year at time of harvest each plat (treated and control) was divided into eastern and western halves, and each section was weighed separately to show any inequalities in soil conditions.

The yields recorded at harvest showed a lack of uniformity in the plats, but gave no evidence of a sensible increase in yield due to the electrical treatment. The results are shown in Table 12.

TABLE 12.—*Yields of rye on plats following electrocultural treatment (alternating charge), section B, Arlington Experiment Farm, in 1915*

Plat	Yields (pounds)		Ratio of treated to control	
	Shock	Grain	Shock	Grain
Eastern half:				
Treated	1,532	565	1.13	1.08
Control	1,350	525		
Western half:				
Treated	1,304	481	.93	.93
Control	1,408	515		
Total:				
Treated	2,836	1,046	1.03	1.01
Control	2,758	1,040		

Experiments in 1916.—In order to measure the relative yielding power of the two plats (treated and control) under normal conditions wheat was again sown in the fall of 1915 and allowed to mature the following summer without electrical treatment of either plat. Table 13 shows the figures recorded at harvest, the north plat being the treated plat of the three preceding years.

TABLE 13.—*Yields of winter wheat on plats without electrocultural treatments, section B, Arlington Experiment Farm, in 1916*

Plat	Yields (pounds)		Ratio of north to south plats	
	Shock	Grain	Shock	Grain
Eastern half:				
North plat	1,568	456.5	0.95	0.84
South plat	1,660	542.0		
Western half:				
North plat	1,448	403.5	.83	.86
South plat	1,752	467.0		
Total:				
North plat	3,016	861.0	.88	.85
South plat	3,412	1,009.0		

SUMMARY OF EXPERIMENTS IN SECTION B

The 1916 results show about 15 per cent difference in the yield of the plats when no electrical treatment was used, the control plat giving the higher yield. During the preceding three years the yields of the two plats were approximately equal. If the 1916 results are accepted as indicating the relative productivity of the two plats under normal conditions, the conclusion follows that during the preceding three years the electrocultural treatment increased the yield 15 per cent or more and that an alternating charge on the network was equally as effective as a high positive charge. During the time the network was connected to the alternating-current power line the charge was changing sign 50 times per second, the maximum gradient was about 1,500 volts per meter, and there was no appreciable ionization at the network. The conditions were so different from those prevailing when the network was charged to a steady high positive potential that it seems highly improbable that the effect on the growing crop would be the same unless the effect is nil under both conditions, the 1916 results not being representative. The latter conclusion seems the more probable, and this is supported by the experiments in section A which follow.

EXPERIMENTS IN SECTION A

A plat in section A of the same dimensions as the one in B was also used for electrocultural tests. The north half of this plat was equipped with a 16-foot network similar to the B network except that it had twice as many cross wires (5 yards apart). The two networks were connected electrically, so that both received the same charge.

Experiments in 1914.—Soybeans were planted in section A in June, 1914, and subjected to a 6,600-volt 25-cycle treatment (alternating charge) continuously from July 15 to October 19, when the crop was harvested. The total weight of the crop from each plat was determined just after cutting, again after drying in the field, and finally after threshing. The weights recorded are shown in Table 14.

TABLE 14.—*Yields of soybeans on plats following electrocultural treatments (alternating charge), section A, Arlington Experiment Farm, in 1914*

Plat	Yields (pounds)			Ratio of treated to control		
	After cutting	After drying	Beans only	After cutting	After drying	Beans only
Treated	4,093	2,776	811.3	0.97	1.13	1.04
Control	4,242	2,446	782.5			

Experiments in 1915.—After the plat had been plowed and put in good shape, rye was seeded on October 22, 1914, and the 6,600-volt treatment (alternating charge) was started November 5 and maintained continuously until harvest. The field was divided into four equal parts when the rye was cut, to get some idea of the soil variation in the eastern and western halves of the plats. At harvest time the crop under the network showed a much better growth than the control plat, but this was probably owing to soil conditions rather than to the electrical treatment, as indicated by the comparative test the following year. The yields obtained are shown in Table 15.

TABLE 15.—*Yields of rye on plats following electrocultural treatments (alternating charge), section A, Arlington Experiment Farm, in 1915*

Plat	Yields (pounds)		Ratio of treated to control	
	Shock	Grain	Shock	Grain
Eastern half:				
Treated	1,270	469	1.46	1.29
Control	868	363		
Western half:				
Treated	1,392	512	1.56	1.52
Control	890	337		
Total:				
Treated	2,662	981	1.51	1.40
Control	1,758	700		

Experiments in 1916.—Rye was again sown in section A in the fall of 1915 and allowed to mature without electric treatment. This crop was cut in June, 1916, giving the yields shown in Table 16, the north plat being the plat treated during the two preceding years.

TABLE 16.—*Yields of rye on plats without electrocultural treatments, section A, Arlington Experiment Farm, in 1916*

Plat	Yields (pounds)		Ratio of north to south	
	Shock	Grain	Shock	Grain
Eastern half:				
North	1,802	557	1.36	1.36
South	1,328	409.5		
Western half:				
North	1,892	590.5	1.54	1.50
South	1,230	393.5		
Total:				
North	3,694	1,147.5	1.44	1.43
South	2,558	803.0		

SUMMARY OF EXPERIMENTS IN SECTION A

A comparison of the yields obtained in the field trials in section A gives no evidence of an increased yield accompanying the use of an alternating charge on the network.

ELECTROCULTURAL EXPERIMENTS IN THE PLANT HOUSE

TRANSPIRATION

The effect of a very high potential gradient on the transpiration rate was investigated in plant-house experiments in Washington in 1913. Large galvanized-iron buckets were filled with moist soil and fitted with special covers to prevent evaporation from the soil. Six rooted geranium cuttings were planted in each pot through holes in the cover, the opening around the stem of the plant being sealed with wax.

The initial weights were taken on February 15, 1913, and the plants were allowed to grow until February 20 without treatment, to determine the relative transpiration of two sets of six pots each. One set was then placed under an insulated frame covered with galvanized-wire screen of ½-inch mesh, while the control set was protected from the discharge by being placed inside a Faraday cage of ½-inch mesh. The frame was connected to the positive pole of the static machine, the other pole being grounded. The frame was charged four hours a day, from 3 to 7 p. m., from February 21 to March 24. The plants were again allowed to grow without treatment from March 25 to April 7. During each period weighings were made to determine the loss due to transpiration, and water was added when necessary to maintain approximately the initial moisture content of the soil.

Table 17 shows the rate of transpiration for each pot during the three periods and the ratio of the treated to the control set. It will be noted that during the period of treatment no sensible change occurred in the transpiration ratio.

TABLE 17.—*Transpiration rate of geranium plants in pots under electrocultural treatment in the plant house at Washington, D. C., in 1913*

Pot designation	Transpiration rate per hour (grams)								
	No treatment		Treatment period					No treatment	
	Feb. 15 to 17	Feb. 17 to 20	Feb. 20 to 25	Feb. 25 to Mar. 1	Mar. 1 to 5	Mar. 5 to 13	Mar. 13 to 24	Mar. 24 to Apr. 1	Apr. 1 to 7
Treated set:									
No. 175	3.6	5.1	5.1	2.9	8.3	8.0	8.8	7.7	10.5
No. 176	3.3	4.7	5.1	3.4	8.4	8.0	9.5	8.2	10.9
No. 177	2.9	5.3	4.8	2.9	8.1	7.6	9.0	7.7	10.9
No. 178	2.6	4.7	4.4	2.9	7.7	7.4	9.4	8.9	11.4
No. 179	4.3	6.1	6.4	3.3	9.7	7.9	7.7	6.8	10.4
No. 180	2.1	5.0	4.9	3.2	7.9	6.8	8.7	8.6	11.4
Mean	3.13	5.15	5.11	3.10	8.35	7.61	8.85	7.98	10.91
Control set:									
No. 181	3.1	4.2	4.3	2.8	7.0	6.5	8.4	9.1	11.1
No. 182	3.3	5.7	6.0	3.7	8.6	8.0	8.6	7.3	10.7
No. 183	3.3	4.6	5.2	3.4	8.1	7.6	8.9	8.2	10.6
No. 184	3.6	5.0	4.9	3.2	8.0	7.6	8.9	8.4	10.9
No. 185	4.3	5.7	5.9	4.0	9.0	8.2	8.1	7.1	10.3
No. 186	3.6	5.3	5.5	3.5	8.7	8.4	8.6	7.4	10.6
Mean	3.53	5.08	5.30	3.43	8.23	7.71	8.58	7.91	10.70
Ratio of treated to control	.89	1.01	.97	.91	1.01	.99	1.03	1.01	1.02

The total transpiration from the treated and control sets of potted geranium plants for the three experimental periods is given in Table 18.

TABLE 18.—*Total transpiration of geranium plants in pots during the three experimental periods in the plant house at Washington, D. C., in 1913*

Designation	Total transpiration (kilograms)		
	No treatment, Feb. 15 to 20	Treatment period, Feb. 21 to Mar. 24	No treatment, Mar. 25 to Apr. 7
Treated set	3.00	33.42	20.01
Control set	3.08	33.42	19.62
Ratio of treated to control	.98	1.00	1.02

WATER REQUIREMENT

An investigation of the effect of a high potential gradient on the water requirement of cowpeas was undertaken in a plant house during the winter of 1918. Eighteen large galvanized-iron cans, each holding about 125 kilograms, were filled with well-mixed soil and fitted with special covers to prevent evaporation. The cowpeas were planted through holes in the covers, the openings being sealed with wax. The pots were weighed at the beginning and at the end of the experiment, and a record was kept of the water added to each pot, from which the total quantity of water transpired by the plants in each pot could be determined. In brief, the procedure was that followed by Briggs and Shantz (*10*, *11*) in their water-requirement measurements.

These pots were divided into three sets of six each. Set No. 1 was placed on an insulated stand, with each pot connected to the positive pole of a static machine; set No. 2 was grounded and placed under a positively charged iron-wire screen suspended about 2 feet above the plants; and set No. 3 was used as a control and was protected from the influence of the charged sets by a well-grounded wire screen. The potential supplied by the static machine was above 50,000 volts.

As soon as the treatment started trouble was experienced with the set beneath the charged network, soot and dust (large ions) being deposited on the leaves and stems of the plants, and in fact all over the house. A coating would collect on the leaves over night during the course of a 16-hour treatment. The plants were washed several times, but they did not thrive, owing in part at least to the great reduction in photosynthesis resulting from the coating on the leaves. This set was finally discarded.

The other two sets, however, grew well throughout the experiment, although they were not so vigorous as plants grown out of doors in the summer. The positions of the pots in a given set were interchanged weekly, so as to provide average light conditions for each pot.

The plants were cut May 2, after 54 days of treatment for 16 hours each day (from 4 p. m. to 8 a. m.), and they were dried at 100° C. and weighed. The water requirement of the plants in each pot was computed by dividing the total weight of water transpired by the dry weight of the crop. The mean water requirement for each set of six pots with its probable error was as follows: For the treated set, 449 ± 4; for the control set, 429 ± 5. A slightly higher water requirement is thus shown for the treated set, the observed increase being 4 ± 1.2 per cent. If some of the water molecules escaping through the stomata of the leaves carried a positive charge, they would move away from the leaf more rapidly than under normal conditions, owing to the strong electric field. This would be equivalent to a virtual increase in the vapor pressure gradient near the leaf and would tend to increase the evaporation rate. Although the above suggestion is highly speculative, it would be of interest to repeat the experiment, applying the electric charge during the daylight hours when the transpiration rate is highest.

SUMMARY OF EXPERIMENTS AT ARLINGTON EXPERIMENT FARM

Electrocultural experiments extending over a period of eight years have been conducted at the Arlington Experiment Farm, Rosslyn, Va., for the purpose of determining whether a highly charged network will increase the yield of crops growing under it. The electrical treatment was usually given during the early-morning and late-afternoon hours. The general experimental procedure was similar to that employed in experiments in England in which the electrical treatment is reported to have given increased yields.

These experiments do not show any well-defined increase in yield due to electrical treatment. There is an indication of a slight increase in the yield of wheat when grown under a positively charged network, but the observed increase is well within the experimental error of field trials.

The results of these field experiments are summarized in Table 19. The relative productivity of the plats when not subjected to the electrical field was determined in order to provide additional information in interpreting the results, a precaution which has not been generally observed by other investigators. A discussion of the yields from each section will be found in the text embodying the description of the experiments.

TABLE 19.—*Summary of the results of the electrocultural experiments in sections A, B, and E, Arlington Experiment Farm, in stated years*

[The treated and control plats in sections A and B were each three-fourths of an acre in area; those in section E half an acre each, separated by an interval of 350 feet. *Abbreviations and symbols.*—Column 2: C=Cowpeas (crop cut for hay); R=Winter rye; S=Soybeans; W=Winter wheat. Column 3: Numbers refer to preceding tables. Column 4: A=25-cycle alternating current; N=No treatment; −=Negative direct current; +=Positive direct current. Column 12: *=Yield of plats treated in previous years]

Section and date	Crop	Table reference	Network treatment: Character of current: Charge	Network treatment: Character of current: Voltage	Network treatment: Description of network: Height (feet)	Network treatment: Description of network: Spacing (yards)	Network treatment: Time of treatment (hours): Per diem	Network treatment: Time of treatment (hours): Total duration	Yields (pounds): Dry shock: Treated	Yields (pounds): Dry shock: Control	Yields (pounds): Grain: Treated	Yields (pounds): Grain: Control	Ratio of treated to control: Dry shock	Ratio of treated to control: Grain
1	2	3	4	5	6	7	8	9	10	11	12	13	14	15
Section A:														
1914	S	14	A	6,600	16	5	-----	----	2,776	2,446	811.3	782.5	1.13	1.04
1915	R	15	A	6,600	16	5	-----	----	2,662	1,758	981	700	1.51	1.40
1916	R	16	N	-------	16	5	-----	----	2,700	2,558	*1,147.5	803	1.44	1.43
Section B:														
1912	W	9	+	45,000	7	1	[1] 16	----	3,465	3,300	1,154	1,114	1.05	1.04
1913	W	10	+	40,000 to 50,000	16	10	[2] 16	----	3,254	3,139	808	782	1.04	1.03
1913	C	----	+	45,000	16	10	[2] 4	128	1,807	1,847	-------	-------	.98	----
1914	Corn	11	A	-------	16	10	-----	----	6,952	6,212	2,892	2,260	1.12	1.28
1915	R	12	A	-------	16	10	-----	----	2,836	2,758	1,046	1,040	1.03	1.01
1916	W	13	N	-------	----	----	-----	----	3,016	3,412	*861	1,009	.88	.85
Section E:														
1913	R	7	N	-------	----	----	-----	----	2,438	2,499	-------	-------	.98	----
1914	W	2	+	30,000 to 60,000	16	5	[2] 4	336	2,332	2,281	644.8	656.5	1.02	.97
1915	W	3	+	30,000 to 60,000	16	2	[3] 6½	345	1,548	1,362	624.5	604.5	1.14	1.03
1916	W	4	−	45,000	16	2	[1] 16	800	2,528	2,444	672	754	1.03	.89
1917	W	5	N	-------	----	----	-----	----	3,190.5	3,064.5	1,137.5	1,198.5	1.04	.95
1918	W [4]	6	+	30,000	16	1	[1] 16	736	2,820	2,639	*1,050	1,025	1.07	1.02

[1] From 4 p. m. to 8 a. m.

[2] From 3 to 7 p. m.

[3] From 4 to 7 a. m. and from 5 to 8.30 p. m.

[4] Plats separated by grounded wire screen.

Plant-house experiments were also made on the effect of an electric charge on the transpiration rate and the water requirement of plants. The effect observed was well within the errors of experiment.

The use of electrocultural methods in their present state of development as a practical means of increasing the yield of crops in this country is not recommended.

REVIEW OF OTHER INVESTIGATIONS IN ELECTROCULTURE

Electrocultural experiments may be divided into two main classes: (1) Those in which the soil is the medium of conduction and (2) those in which the air is the medium of conduction. Experiments of the first class cover the use of soil currents resulting (1) from an externally applied electromotive force, (2) from the galvanic action of the soil moisture on zinc and copper plates buried in the ground, and (3) from the use of metallic uprights designed to collect and carry atmospheric electricity to the soil. Experiments of the second class are those in which the normal air-earth current is increased by means of a highly charged network over the plants or decreased by inclosing the plants in a grounded cage made of metal screen.

EXPERIMENTS WITH SOIL CURRENTS

Among the first experiments with soil currents on a large scale were those by Ross, prior to 1844, (*44*) in New York. He buried a copper plate 5 feet by 14 inches perpendicularly in the earth with the 5-foot edge horizontal, and at a distance of 200 feet a zinc plate of the same dimensions was similarly buried. The two plates were connected above the ground, forming a galvanic cell. Potatoes were drilled in rows between the plates and also in a similar plat without plates. At the end of the experiment some of the potatoes from both plats were measured, those from the treated plat averaging 2½ inches in diameter, while those from the control averaged only half an inch. The total weights at harvest are not given, and conclusive assurance that the two areas were of equal fertility at the outset is lacking. The supposed beneficial effect is rendered doubtful through the subsequent discontinuance of so simple a treatment.

About this time Solly (*46*) conducted in England 70 small tests similar in principle to those of Ross, the plates being 4 by 5 inches and spaced only 6 inches apart. Grains, vegetables, and flowers were planted between the electrodes. On comparing the appearance of the treated and untreated plants a beneficial effect was recorded in 19 cases, a harmful effect in 16 cases, and no effect in 35 cases. Solly concluded that electricity has practically no effect on plant growth.

Fitchner (*16*) has recorded large increases from treatment with galvanic currents. From his figures alone the experiments would indicate increases of 16 to 127 per cent due to treatment. The statement was made, however, that the treated plats were provided with drains but that the control plats were not. Such conditions do not constitute good experimental practice and leave the results open to question. This same objection holds for accompanying experiments on the decomposing action of the galvanic current on soil.

62149°—26——3

In 1881, F. Elfving (*15*) undertook an interesting series of experiments with different seedlings growing in culture solutions through which he passed battery currents of different strengths. After germination the seedlings were mounted on corks which were floated in the solution between electrodes 6 by 4 centimeters in size. He found that in nearly every case the root would turn and grow in a direction against that of the electric current. Plates of carbon, zinc, and platinum were used, and all gave the same effect. Elfving attributes this phenomenon of orientation to the slowing up of the growth on the side of the root toward the positive pole. This same phenomenon was noticed by Plowman (*40, 41*) in 1902–03.

Holdefleiss (*23*) in 1884 selected several rows of sugar beets in a field which showed a good stand and uniform conditions. In this field copper plates 50 centimeters square were sunk perpendicularly in the ground 50 centimeters deep, so that the plates included two rows of beets. At the other end of the rows, 56 meters distant, other plates were sunk, and between the two a 14-cell Meidinger battery was connected. This same arrangement was used on a potato field. Further experiments were conducted with copper and zinc plates 33 meters apart connected by a solid copper wire. The report of the experiments stated, in substance:

(1) That an electric current was present on all treated plats throughout the season, its presence being determined by a sensitive electrometer; (2) that the rows of beets and potatoes between plates which were connected to the battery showed no difference in growth at any stage of their development; (3) that the beets and potatoes in rows between the zinc-copper combinations assumed a somewhat fresher and stronger appearance about 10 days after the beginning of the experiment, and the harvest showed an increased yield ranging from 15 to 24 per cent.

It should be remembered, however, that in experiments with soil currents the path of the current is not wholly by the most direct route from one electrode to the other, but that the lines of flow spread out through the soil in a way similar to the spreading of the lines of force between the poles of a bar magnet.

Experiments conducted by Wollny (*48*) included five plats 4 by 1 meter each in size separated by a path 1.2 meters wide and by boards sunk 25 centimeters in the ground. On plats 1 to 3 a zinc plate was sunk at both of the narrow sides, and these were connected as follows: Plat 1, induction coil operated by three Meidinger elements; plat 2, a battery of six Meidinger elements; plat 3, a battery of three Meidinger elements. On plat 4 a zinc plate was sunk on one end and a copper plate at the other, the two being connected above ground by a copper wire. Plat 5 constituted a check or control plat. Each plat was divided into four equal parts 1 square meter each in area and seeded. Numbers of plants up on different dates showed practically no effect for any of the different treatments. The yields recorded at harvest time, based on an equal number of plants per square meter, are shown in Table 20.

TABLE 20.—*Yields of rye, rape, bean, and potato plants after electrocultural treatments in 1883, according to Wollny*

Plat	Treatment	Yields per square meter (grams)			
		Rye, 42 plants	Rape, 42 plants	Beans, 42 plants	Potatoes, 5 plants
No. 1	Induction	182.0	114.8	517.5	372.1
No. 2	6 cells	219.8	94.5	514.5	310.5
No. 3	3 cells	197.8	103.0	420.0	315.3
No. 4	Cu-Zn	201.6	114.7	600.0	397.8
No. 5	Control	228.7	118.7	631.0	377.6

These records show that in nearly all cases the control plat gave the best yields, but further experiments were conducted in 1886 and 1887. The ground was well worked over, and four plats 16 by 2 meters were selected, separated from each other by paths 1.2 meters wide and bordered by wooden lath walls. Each plat was divided into eight smaller plats 2 meters square and all were given equal applications of manure. On the small ends of the four large plats zinc plates 2 meters by 30 centimeters in area were sunk perpendicularly and connected above ground through an induction coil operated by 4 or 5 cells for plat 1 and through a 4 or 5 cell battery for plat 2. Plat 3 served as a control, and plat 4 had a copper plate at one end directly connected by a copper wire with a zinc plate at the other end. Diagonally lying plats were planted with the same crops, the grains being drilled to give a uniform planting. The presence of a current on all treated plats was noted by the use of a galvanometer. Throughout the season there was no perceptible difference in growth between treated and control plats during either year. The comparative-yield weights are shown in Table 21.

TABLE 21.—*Yields of vegetable crops after electrocultural treatments in 1886 and 1887, according to Wollny*

Plat	Treatment	Yields per plat 2 meters square (grams)							
		Rye	Rape	Peas	Beans	Corn	Potatoes	Beets	Turnips
In 1886:									
No. 1	Induction	113.3	339.0	1,420.0	2,040.0		6,400	23,400	22,250
No. 2	5 cells	108.6	300.5	1,570.0	2,410.0		4,650	24,420	18,080
No. 3	Control	107.8	405.8	1,380.0	2,220.0		6,620	28,100	21,520
No. 4	Cu-Zn	100.9	418.0	1,490.0	2,190.0		6,670	29,400	20,800
In 1887:									
No. 1	Induction	933.0	775.0	548.0	696.5	1,962.8	8,350	19,640	17,850
No. 2	4 cells	879.0	755.0	588.0	607.0	1,923.6	8,190	17,650	18,270
No. 3	Control	948.4	773.0	592.0	584.2	1,913.6	8,410	18,900	18,460
No. 4	Cu-Zn	838.5	761.6	571.0	465.0	2,072.9	8,920	16,320	19,660

From these experiments Wollny concluded that an electrical current conducted through soil in which plants were growing had in general no influence or possibly a harmful effect on the productiveness of the plants.

Leicester (*29, 30*) used boxes of soil 2½ by 3 feet in area, with copper and zinc plates connected above ground. Control boxes without plates were included. After several trials with different kinds of seeds, it was found that in every case the seeds grew much quicker in the boxes containing the plate. Hemp seed was fully an inch above the surface before controls showed any plants. The observation was made also that plants in the zones nearest the plates were the first to come up. Watering with dilute acetic acid was found to cause quicker growth for treated plants—possibly because of increased current resulting from the acid-metal reaction. Upon repeating these experiments, Leicester decided that the only action of the current was to stimulate the plant until the initial store of food was used up. No data were recorded in either of his reports.

Berthelot (*3*) conducted some tests with soil currents to determine whether electricity aided in the fixation of nitrogen by plants. Suitable control plats were provided. He reported that the treated plants grew much more rapidly, being nearly twice the weight of the control plants at the end of four to six weeks. Although not complete or definite, the experiments were abandoned for various reasons.

Kinney (*27*) made an extensive series of experiments to determine the influence of electrical currents on germination. Seeds were subjected to different current strengths for different periods of time and then put in suitable germination apparatus and the subsequent growth noted. An intermittent treatment of 30 seconds per hour was given in some cases, arranged by clock contacts. Two different arrangements were used for the treatments. In one a glass cylinder containing the seeds was equipped at each end with electrodes. These were pressed against the seeds through which the current was thus directly passed. In the other, the seeds were placed in wet sand held between perforated metal disks, which were used for the electrodes. The entire layer was held in a glass funnel in which the growth of the radicle could be measured without removal. Eight sets of 25 seeds each were used in each test, one set being the control and the other seven receiving different strengths of current. Experiments with barley showed that the growth of treated seeds increased as the current strength increased up to a certain optimum value, above which the growth decreased with increase in current strength. With white mustard, rape, and red clover the optimum treatment for both roots and stems was identical.

Plowman (*40*, *41*) has recorded the results of experiments conducted at the Harvard Botanical Gardens on the influence of soil-conducted currents on plant life. Platinum or carbon electrodes were used, with potentials ranging from 5 to 500 volts. The regulation of temperature was a serious difficulty—a fact mentioned for the first time in connection with such experiments and one that may have been ignored in earlier reports. Plowman found that seeds near the anode were always killed by a current of 0.003 ampere or more if continued for 20 hours. Seeds at the cathode were little affected by currents less than 0.08 ampere.

Gerlach and Erlwein (*19*, *20*), at Bromberg, investigated the effect of weak soil currents on germination and growth. The field was made up of seven plats of 200 square meters each. Current was taken from a car line and led to the three treated plats, which were provided with iron plates 20 meters long by 30 centimeters wide and 2 millimeters thick sunk into the soil at both ends. Each of the seven plats was seeded half with barley and half with potatoes.

The treatment continued 24 hours a day for 86 days for barley and 139 days for potatoes, beginning in April. Both barley and potatoes showed excellent growth, but no differences between the treated and control plats were discernible at any time. Other experiments were conducted with plants grown in boxes provided with copper and zinc plates connected overhead by wires. Trials with rye, wheat, and lupine gave no difference between treated and untreated crops.

Homberger (*24*) reported that the passage of high-frequency currents through the soil was beneficial to plant growth. His experiments were conducted on a small scale, using flowerpots with only a few plants, the treatment consisting of three applications daily until the temperature of the soil reached 35° C., when the current was cut off. The leaves and stems of the treated plants showed more chlorophyll than the controls. A photograph shows one pot each of treated and control plants, the treated plants being about five times as high as the others. In order to determine whether the heating was the main cause of increased growth another pot was subjected to test currents for five minutes daily. These plants were about four times the height of the controls when photographed. From these comparisons Homberger concluded that the oscillating field and not the temperature was the main cause of the stimulation, and he believed his results to be due to chemical changes taking place under the influence of the oscillating electromagnetic field, analogous to the catalytic action of light.

In 1907 (*17*) and 1909 (*18*) Gassner reported upon experiments with charged soil which indicated a general unfavorable action upon plant growth.

Kövessi (*28*) obtained unfavorable results in researches involving some 1,100 experiments.

Considerable publicity has been given to an apparatus called a "geomagnetifier," a sort of lightning rod designed to gather in atmospheric electrical energy and supply it to the crops. Among those who have reported favorable results through the use of such apparatus are Maccagno (*35*), Basty (*2*), and Paulin (*39*).

At the present time methods of electroculture employing soil-conducted currents have few proponents.

EXPERIMENTS WITH MODIFIED POTENTIAL GRADIENTS

Grandeau (*21*), in 1878, reported studies on the effect of the electrical condition of the atmosphere upon the growth of vegetation He grew plants in a Faraday cage consisting of four iron rods 1. centimeter in diameter by 1.5 meters high, holding fine iron wires forming 15 by 10 centimeter meshes. The cage was grounded in order to destroy the normal electrical field. Experiments were made with tobacco, corn, and wheat. The plants under the cage were reported weak and slender. Six stalks of wheat grown in free air weighed 6.57 grams, as compared with 4.95 grams for six stalks grown under the cage.

Grandeau was led by these experiments to the belief that high trees act as a grounded network, in that they shield the vegetation beneath their foliage from the action of the normal electrical field, thereby causing a decreased rate of growth. With a sensitive Thompson electrometer, he compared the strength of the field in the open with that under vegetation. The results indicated that under trees and shrubs the potential gradient was greatly reduced. The experiments of Grandeau were confirmed by Mascart (*36*).

As opposed to the conclusion of Grandeau, the modern greenhouse of steel construction constitutes in itself an approximation to a Faraday cage about the plants growing within it, and yet the development of the plants is surely not seriously impaired in consequence. Likewise, Briggs and Shantz (*10, 11*), in their investigation of the water requirements of plants, carried hundreds of pots of plants to full maturity under a grounded metal framework, covered above and on the sides with metal screen of ½-inch mesh, which must have annulled the normal electrostatic field; yet the plants grown within the inclosure were almost without exception superior in development and luxuriance of foilage to those grown in similar pots outside.

Lemström (*32*) conducted in Finland a long series of experiments to determine, if possible, the influence of static electricity on plant growth. The presence of strong electric charges in the atmosphere of northern regions, as indicated by the northern lights, linked with the astonishing development of vegetation in such regions, led him to regard atmospheric electricity as an important factor in plant growth. Garden vegetables, fruits, and small grains were subjected to several different treatments in these investigations both in greenhouses and in open fields. Lemström summarized the results of his experiments as follows:

(1) The real increase due to electrical treatment has not yet been exactly determined for the different plants, but we are approaching its smallest value by fixing it at 45 per cent.

(2) The better and more scientifically a field is cultivated and manured, the greater is the increase percentage. On poor soil it is so small as to be scarcely perceptible.

(3) Some vegetables can not endure the electric treatment if they are not watered, but then they will give very high percentage increases. Among these are peas, carrots, and cabbage.

(4) Electric treatment when accompanied by hot sunshine is damaging to most vegetables, probably to all; wherefore if favorable results are to be arrived at the treatment must be interrupted in the middle of hot and sunny days.

Experiments similar to those conducted in Finland were conducted in England, Germany, and Sweden with like results. A detailed description of all of these experiments may be found in "Electricity in Agriculture and Horticulture," by Lemström (*32*).

Priestley (*42, 43*) reported on the experiments of Newman (*37*) at Golden Valley Nurseries at Bitton. A small Wimshurst machine was used, one terminal of which was grounded and the other connected to wires suspended over outside plats and also to wires in seven glasshouses. The wires were hung 16 inches above the tops of the plants and were provided with discharge points hung at short intervals. The machine was operated 9.3 hours a day for 108 days between March 27 and July 26, the first half of the period in daytime and the latter half at night. Control plats were provided in all cases similar to the treated plats except without wires. The results recorded are given in Table 22.

TABLE 22.—*Results of electrochemical treatment of garden crops at Bitton, as reported by Newman*

Crop	Treated plants	Notes
	Per cent	
Cucumbers, increase	17	Less subject to bacterial disease.
Strawberries:		
5-year plants, increase	36	
1-year plants, increase	80	More runners produced.
Broad beans, decrease	15	5 days earlier.
Cabbage	----------	
Celery, increase	2	10 days earlier.
Tomatoes (no difference)	----------	

During the same year an installation was working at Gloucester with higher voltage and wires 5 feet from the ground. The following results with treated plants were reported: Beets, 33 per cent increase and higher total sugar content; carrots, 50 per cent increase; turnips showed an increase, but the percentage was not recorded owing to slugs.

In 1906 Newman (*37*) and Lodge (*33*), at Evesham, began some electroculture experiments using about 40 acres, 20 of which were electrified with a network 15 feet above ground. The Lodge apparatus was used, 22 poles carrying the wire over the area, with small wires 12 yards apart. These experiments were continued several years. The results are summarized in Table 23.

TABLE 23.—*Results of electrochemical treatment of crops at Evesham in stated years, as reported by Newman*

Year and crop	Electrified crops	Notes
1906:		
Wheat (electrified area 12 acres)—	*Per cent*	
Canadian, increase	39	Sold for 7½ per cent higher price when bakers found it produced a better baking flour. The somewhat poor yield from the control plat was probably due to deficiency in lime, afterwards rectified.
English, increase	29	
1907:		
Wheat (electrified area 11 acres), increase.	29 18	Estimated by cartloads.
Strawberries, increase	25	
1908:		
Wheat (electrified area 7.68 acres), increase.	24.3	
Strawberries, decrease	9	Dry season.
Tomatoes, increase	30	By weight per plant (average).
Cucumbers, increase	8.4	By number (average).

Newman reported later (*38*) that during seven successive years (1905 to 1911) wheat gave an average increase of 21 per cent in weight of grain and an increase of straw which it was not possible to measure.

Potato variety experiments conducted at Dumfries, Scotland, by Dudgeon in 1911 and 1912 (*14*) gave the yields shown in Table 24.

TABLE 24.—*Results of electrocultural treatment of potato varieties at Dumfries, Scotland, by Dudgeon in 1911 and 1912*

Variety	Yield (tons)		Variety	Yield (tons)	
	Treated	Control		Treated	Control
Ringleader	8.05	5.85	Golden Wonder	8.74	8.12
Windsor Castle	11.72	9.88	Great Scot	11.79	10.31

In 1912 further experiments at Dumfries were carried on in another field, exposed to wind from any quarter. Two corners of the 4 acres were treated, the others left as controls. No difference in yield was recorded, and it is explained that probably all plats are to be regarded as treated plats.

In 1915 Dudgeon conducted an experiment with oats. The crop was grown on ground that had been used for similar experiments on potatoes for three years. Two adjacent plats of 1½ acres each were separated by a well-grounded wire screen 3 feet higher than the charged network. A sensitive electrometer showed that the screen reduced the leakage over the control plat but did not altogether prevent it. The season was dry and the crop was not heavy. From early stages the treated plat showed a marked superiority in comparison with the control, and did not suffer from the prevailing drought to the same extent. The electrical discharge was applied about five hours each day for 108 days. The weights (pounds) recorded at harvest were as follows: Treated—grain, 1,309, straw 2,476; control—grain 1,008, straw 1,572.

These figures indicate an increase of about 30 per cent in grain and about 58 per cent in straw. Analyses of the grain from the two plats showed practically no difference in quality.

Blackman and Jørgensen (*6*) have also reported experiments by Dudgeon at Dumfries, Scotland, with oats. In a 9-acre field 1 acre was selected for treatment and two half-acre plats for controls. The distance between the silicon-bronze wires of the network was 4.5 yards. Current of 3 amperes at 50 volts was supplied to the primary circuit, giving a greater intensity of discharge than that obtained in the experiments of the previous years. The discharge was started just as soon as the crop appeared above ground, and within a month a marked difference was noted. The treated plants had deeper color and were higher than the control plants. Throughout the season the treated crop was 5 to 10 inches higher than the control. Plants around the network also showed the effect of the discharge. The total application from April 14 to August 17, daytime only, was 848 hours. Heavy rains did a good deal of damage. The comparative yields were as shown in Table 25.

TABLE 25.—*Results of electrochemical treatment of oats at Dumfries, Scotland, by Dudgeon, as reported by Blackman and Jørgensen*

Field	Yields (pounds)				
	Grain		Straw		
	Quality 1	Quality 2	Bunches	Total	Per bunch
Control 1 (half acre)	630	210	99	1,218	12.3
Treated (acre)	1,942	695	316	4,924	15.6
Control 2 (half acre)	714	210	103	1,401	13.6

These results indicate a 49 per cent increase in grain and an 88 per cent increase in straw for the electrical treatment.

The Liverpool City and Electrical Engineers reported on experiments conducted near Liverpool, England, in 1917. Two plats in newly plowed pasture land separated by about 375 feet were used, an analysis indicating that the surface and subsoil were of the same character. Various plant crops were grown, and in general the electrified area gave substantial increases in yield over the control area. A copy of this report is on file in the Office of Biophysical Investigations, Bureau of Plant Industry.

Honcamp (*25*) has summarized the results of several previous investigations and pointed out serious objections to the methods used.

TABLE 26.—*Results of electrochemical treatments of oat crops at Mocheln, Germany, according to Gerlach and Erlwein*

Electrical and soil treatment	Relative yields		Composition (per cent)			
			Grain		Straw	
	Grain	Straw	Dry matter	Nitrogen	Dry matter	Nitrogen
No electricity:						
Fertilizer, irrigation	26.60	34.40	91.4	1.94	77.4	0.32
Do	28.80	32.20	92.4	1.70	76.2	.27
Fertilizer, no irrigation	20.60	24.40	87.9	2.13	73.5	.46
Do	20.90	22.10	89.9	2.04	74.3	.38
No fertilizer, no irrigation	19.60	19.40	90.4	1.85	78.8	.32
Do	19.60	17.40	90.1	1.67	79.8	.32
Direct current:						
Positive, fertilizer, irrigation	27.80	36.20	90.4	1.94	71.6	.27
Negative, fertilizer, irrigation	27.80	37.20	91.7	1.74	72.9	.24
Positive, fertilizer, no irrigation	21.60	24.40	90.5	2.16	74.5	.36
Negative, fertilizer, no irrigation	20.50	22.50	91.2	2.07	73.7	.30
Positive, no fertilizer, no irrigation	21.00	23.00	84.4	1.88	76.6	.28
Negative, no fertilizer, no irrigation	17.80	18.20	90.4	1.72	78.4	.28
Alternating current:						
Fertilizer, irrigation	26.20	31.80	91.7	1.81	78.4	.28
Do	26.00	32.00	90.6	1.78	77.9	.23
Fertilizer, no irrigation	19.50	21.50	91.3	2.16	73.2	.33
Do	19.90	21.10	89.8	2.11	75.9	.30
No fertilizer, no irrigation	18.00	18.00	91.1	1.84	82.2	.28
Do	18.00	18.00	91.4	1.83	85.0	.26

SUMMARY OF RELATIVE YIELDS OF GRAIN AND STRAW

Soil treatment	No electricity	High-tension current		
		Direct		Alternating
		Positive	Negative	
Grain:				
No fertilizer, no irrigation	19.60	21.00	17.80	18.00
Fertilizer, no irrigation	20.75	21.60	20.50	19.70
Fertilizer, irrigation	27.70	27.80	27.80	26.10
Straw:				
No fertilizer, no irrigation	18.40	23.00	18.20	18.00
Fertilizer, no irrigation	23.25	24.40	22.50	21.30
Fertilizer, irrigation	33.30	36.20	37.20	31.90

On the Continent during this period many electrocultural experiments were carried out, using networks charged to high potentials. Reports by Höstermann (*22*), Gerlach and Erlwein (*19, 20*), Clausen (*13*), Breslauer (*9*), and others indicate that no benefit may be expected from the use of the network. The German experiments made use of an extensive and variable complex of conditions, designed to include the study of positive and negative potential in relation to fertilizers and irrigation and the relation of these factors to the composition of grain and straw. The results shown by Gerlach and Erlwein reporting experiments with oat crops at Mocheln are selected as representative. (Table 26.)

It may be well worth while to consider Table 26 in some detail, since it seems to represent a thoroughly impartial study of the methods which have given success elsewhere.

The instances in which duplicate trials were run and the agreements to be noted for these cases show rather conclusively that lack of uniformity in soil conditions was not a disturbing factor in these experiments. The six plats giving notably higher yields are those with fertilizer and irrigation. These are in good agreement and show no appreciable advantage for the three types of electrical treatment represented, the averages for relative yields only being as shown in the summary of Table 26.

The plats in these experiments were about one-fourth acre each, the control plats being separated from the electrified plats by about 325 feet. The potential of the direct-current network was about 30,000 volts, whereas that of the alternating current was about 20,000 volts. The statement of Lemström that the better the condition of the field the more favorable the influence of the high-tension discharge is not substantiated by these trials. In brief, the German experiments give little evidence of any definite crop increase attribntable to the electrical treatment.

In 1913 Dorsey conducted greenhouse experiments in Ohio with radishes and lettuce, using a high-frequency current. In a letter to Doctor Briggs dated August 18, 1913, he reported the relative weights of 10 plants selected at random from each area. These are shown in Table 27.

TABLE 27.—*Results of electrocultural treatments of greenhouse radishes and lettuce in 1913, according to Dorsey*

10 plants	Relative weights (grams)			
	Radishes		Lettuce	
	Treated	Control	Treated	Control
Total	265.7	180.0	67.0	46.1
Edible portion	139.5	79.4	60.7	41.8
Tops	120.5	95.0		
Roots	9.3	5.6	6.3	4.3

Dorsey also conducted field trials with a high-frequency current. The plants used were beets, lettuce, cabbage, beans, melons, cucumbers, and tobacco. They were planted in long rows, one-half of each row being under the charged network. The treated plat covered about half an acre. The network was 9 feet above ground with wires 15 feet apart and carried a voltage of about 50,000 at an estimated frequency of about 30,000 cycles. The power was taken from a 7½ kilowatt 220-volt transformer supplying 11,000 volts at 60 acycles and exciting an oscillating circuit containing the network as capacity. Treatment was given daily, three hours in the forenoon and three hours in the afternoon. A generally favorable influence for the discharge treatment was reported. Unfortunately total weights were not included. The results for the second year were generally unfavorable for the discharge treatment, and Dorsey concluded that perhaps slight differences in the slope of the two plats may have been responsible for the favorable results of the first year.[5]

At the present time perhaps the best evidence of plant response to electrical discharge is that obtained by Blackman (*4, 5, 6, 7, 8*) of the electroculture committee of the British Ministry of Agriculture and Fisheries. His experiments extend over a period of years and comprise field trials, pot cultures, and laboratory tests, all of which he interprets as affording converging evidence for a favorable growth response to the application of electricity. On account of the practical possibilities associated with a treatment assuring increased growth it seems desirable to examine in some detail the data which have given rise to this assurance.

The field trials carried on in England by Blackman and his associates have given the results which are summarized in Table 28.

TABLE 28.—*Results of electrocultural treatments of grain crops in England, as reported in field experiments by Blackman*

Crop and year	Location	Acreage		Duration of treatment (hours)	Yield per acre (bushels)		Ratio of treated to control
		Treated	Control		Treated	Control	
Oat:							
1915	Lincluden	1.5	1.5	557	20.7	16.0	1.29
1916	do	1.0	1.0	848	62.8	42.0	1.49
1917	do	.33	.33	1,060	54.8	48.9	1.12
1917	do	.33	.33	1,060	42.2	44.9	.93
1917	do	.33	.33	1,060	36.9	38.1	.96
1918	do	.25	.06	704	75.5	56.1	1.34
1918	do	.25	.06	704	84.9	58.4	1.45
1918	do	.25	.06	704	80.4	46.3	1.73
1919	do	.11	.11	710	36.6	45.2	.80
1919	do	.11	.11	710	45.1	43.8	1.02
1919	do	.11	.11	710	53.3	28.9	1.84
1919	Harper Adams Agricultural College.	.50	.33	456	47.0	53.6	.87
1919	do	.25	.33	456	63.8	48.2	1.32
1919	do	.25		456	50.8	48.2	1.05
1919	do	.50	.33	456	60.2	59.6	1.01
1920	Lincluden	.11	.11	911	36.2	44.8	.80
1920	do	.11	.11	911	43.5	46.1	.94
1920	do	.11	.11	911	51.8	33.0	1.56
1920	Harper Adams Agricultural College.	.33	.33 ?	793	50.0	56.0	.89
1920	do	.33		793	52.5	56.0	.93
Barley:							
1917	Rothamsted	.0125	.0125	1,500	17.8	13.1	1.35
1918	do	.66	.10	643	44.7	36.4	1.22
1918	do	.66	.10	643	47.4	52.7	.89
1918	do	.66	.10	643	40.4	36.3	1.11
1920	do	.50	.50	786	31.7	29.5	1.07
1920	do	.50	.50	786	33.0	25.17	1.31
Winter wheat:							
1919	do	.50	.50	854	21.4	14.3	1.49
1919	do	.50	.50	854	22.3	17.4	1.28
1920	do	.25	.25	727	18.84	20.4	.92
1920	do	.25	.25	727	18.35	18.24	1.006
Spring wheat:							
1919	do	.50	.33	940	7.6	10.0	.76
1919	do		.33	940	7.3	7.9	.92
1919	do	.50	.33	940		6.3	1.15
Average							1.14

[5] Correspondence with the Office of Biophysical Investigations, Bureau of Plant Industry, September 2, 1924.

These tabulated values are in many ways not subject to biometrical analysis; they represent the results of experiments carried out with varied complexes of soil, season, acreage, crop, and electrical treatment. Nevertheless, in the absence of any definite knowledge concerning the conditions under which an electrical treatment may be presumed to be most effective there is perhaps no better index than a comparison.

Of 33 trials shown in Table 28, 21 indicate an increase for treated areas, whereas 12 indicate a decrease. The treated areas return a yield represented by the range 76 to 184 when the untreated areas return a yield represented by 100 and give an average increase of 14 per cent. This increase is based upon yields reported for experiments regardless of crop or seasonal normality, and Blackman estimates the more reliable experiments as indicative of an average increase in yield of about 22 per cent. In either case, such an increase would seem sufficient to be of promise from an agricultural standpoint. If an attempt is made to determine from these tabulated values the conditions under which the increases were obtained, serious difficulties are immediately encountered.

Unfortunately the normal productivity of the electrified and control areas is in most cases unknown, and a serious lack of soil uniformity is evident from the yields of different portions of control areas. For example, in the 1919 and 1920 plats with oats at Lincluden, which occupied the same areas for the two years, the control yields were as shown in Table 29, in which the relative yields of the corresponding treated areas for the same years, the controls being taken as 100, are also shown for comparison:

TABLE 29.—*Comparison of the results of electrocultural experiments with oat crops at Lincluden, England, in 1919 and 1920*

Area	Acre yields of control plats (bushels)		Relative yields of electrified areas, the controls being taken as 100	
	1919	1920	1919	1920
Section I	45.2	44.8	80	80
II	43.8	46.1	102	94
III	28.9	33.0	184	156

It is obvious that the yields of the third section of the control area were uniformly low compared with the yields of the other control sections and that this fact is almost certainly involved in the high percentage increases arising for the third section of the treated area. It would therefore appear that these particular increases may be attributed to a lack of soil uniformity, and the importance of this unknown factor is indicated.

The most consistent series indicating favorable response to electrical treatment appears to be the 1918 oat trials at Lincluden. The plats in oats at Lincluden gave the average annual yields shown in Table 30.

The yields from the electrified areas in 1918 seem to have been so exceptional compared with the electrified areas for the three other years that one may question whether it is justifiable to attribute the increase solely to the electrical treatment.

TABLE 30.—*Analysis of the average results of electrical treatments of oat plats at Lincluden, England, in the years 1917 to 1920, inclusive*

Year	Average acre yields (bushels)		Ratio of treated to control	Year	Average acre yields (bushels)		Ratio of treated to control
	Treated	Control			Treated	Control	
1917	44.6	44.0	1.01	1919	45.0	39.3	1.14
1918	80.2	53.6	1.49	1920	43.8	41.3	1.06

The instances specifically considered in Tables 29 and 30 comprise the most notable of the percentage increases reported by Blackman, as shown in Table 28, and they are therefore in a large measure the basis of the 22 per cent average increase reported. One is thus left without definite assurance that the field experiments demonstrate a favorable response to the electrical treatment.

The pot-culture experiments in England by Blackman and his associates gave results which are summarized in Table 31.

TABLE 31.—*Results of electrical pot-culture experiments with grain crops in England, according to Blackman*

Year and crop	Yields (grams)		Ratio of treated to control
	Treated	Control	
1918:			
Wheat	0.72	0.73	0.98
	.71		.97
Maize	1.43		1.02
	1.21	1.39	.87
	1.24		.89
Barley	2.26		.98
	2.12	2.30	.92
	1.98		.86
	.97		.75
	1.17	1.29	.90
	1.12		.86
Maize	8.12		1.04
	8.37	7.78	1.07
	7.41		.95
	10.84		1.26
	10.36	8.54	1.21
	10.85		1.27
1919:			
Maize	5.70	5.62	1.01
	6.81	5.67	1.20
	2.52	2.32	1.08
	16.73		.96
	16.03	17.28	.92
	15.01		.86
Barley	13.72		.73
	15.84	18.69	.84
	11.75	16.89	.69
1920:			
Maize	15.66	14.52	1.07
	15.39		1.05
Barley	23.84		.99
	23.28	23.88	.97
	26.33		1.10
Wheat	17.66		1.08
	17.11	16.22	1.05
	18.76		1.15
1921:			
Barley	10.60	10.29	1.03
	11.79		1.14
	34.0	31.5	1.07
	37.2		1.18
	49.2		1.05
	53.0	46.5	1.13
	48.5		1.04
	51.9		1.11
	22.8	21.2	1.07
Maize	15.5	15.6	.99
	14.27		.91
	18.20	10.00	1.09
	18.95		1.14
Average			1.01

As with the tabulated values for field experiments, so here also the results of the pot-culture trials represent more than the electric discharge variable; soil and seasonal factors vary as well as the crop and the duration, nature, and strength of the electrical treatment. Making a comparison, 26 trials out of 47 give positive results, while 21 give negative results. The treated plants return yields represented by the range 73 to 127: when the untreated plants return a yield represented by 100 and give an average increase of 1 per cent. This increase is well within the experimental error, and the pot-culture trials in their entirety thus furnish no definite evidence of a response to the electrical treatment.

In contrast to the field experiments, however, the pot-culture trials afford results from several similarly treated pots and plants, so that an estimate of individual experiments may be made by comparing the differences between treated and untreated plants with the probable errors involved in the measurements.

When the pot-culture records are examined in this way, it becomes evident that the treated and untreated plants present substantial differences. With uniform soil and seasonal factors for electrified and control plants the association of these differences with the treatment becomes intimate. The fact that these differences favor the control plants about as often as the treated plants emphasizes the complexities involved and makes one less certain that these differences are definitely attributable to the electric discharge.

The laboratory experiments of Blackman and his associates have been on the effect of a direct current of very low intensity on the rate of growth of the coleoptile of barley. Differences in the growth rate of treated and control plants were noted over short periods. The small differences attributable to the direction of the current and the pronounced after effects obtained make the interpretation of the data difficult and uncertain.

In general, then, one finds in Blackman's experiments many significant differences between the electrified and control plants. In some instances the relation of the discharge to these differences may well be questioned. In others the relation appears to be an intimate one, and the significance of such differences is the immediate concern of further research in electroculture.

TABLE 32.—*Summary of electrocultural trials*

Method	Definite influence reported		No definite influence reported	
	Year	Observer	Year	Observer
Soil-conducted currents:				
Germination	1889	Warren	1893	Bruttini.
	1892	Leicester	1899	Ahlfvengren.
	1897	Kinney	1902	Flammarion.
	1902	Plowman	1905	Löwenberg.
Pot cultures	1892	Warner	1846	Solly.
	1903	Stone		
	1914	Homberger		
Field trials	1860	Fitchner	1883–87	Wollny.
	1884	Holdefleiss	1907	Gassner.
	1844	Ross	1909	Gerlach and Erlwein.
Soluble plant food	1916	Yegoron		
Modified atmospheric potential gradient:				
Increased potential	1904	Lemström		
	1905–1911	Lodge, Newman	1907	Gassner.
	1913	Dorsey	1909	Gerlach and Erlwein.
	1914	Jørgensen, Priestley, Dudgeon.	1909–1910	Breslauer.
			1910	Höstermann.
			1911	Clausen.
	1917	Blackman, Liverpool engineers.	1918–1924	Briggs, Campbell, Heald, Flint.
Decreased potential	1876	Mascart	1880	Laikewicz.
	1878	Grandeau	1914	Briggs and Shantz.
	1910	Höstermann		

A review of the literature of electrocultural experimentation up to the present time does not lend assurance of great progress. (Table 32.) In 1800 Senebier (*45*) wrote substantially as follows:

The researches of Maimbray, Nollet, Bose, Menon, and Jalabert would indicate that electricity accelerated the development of plants, both in their germination and in their subsequent development. Nuneberg, many years afterward, repeated the same experiments with the same results. Linné and Kostling observed the same effects. Achard confirmed these results. Berthelon, in a treatise on the electricity of plants, has summarized the information on the subject and substantiated it by further research of his own. Gardini, from work carried on at Lyon, affirmed the influence of electricity on vegetation. Carmoy, d'Ornoy, and Rosieres have defended this opinion in the Journal de Physique. These doctors base their conclusions on the identity of natural and artificial electricity, on the continual electrified condition of the atmosphere, and on the meteorological phenomena which indicate in a more or less sensitive manner the presence of electricity; the different elevated parts of plants, which are in themselves excellent conductors of electricity, offer in their leaves, as De Saussure has observed, the proper points to receive the electric fluid. . . . All these experiences led to the opinion stated when Ingenhousz published experiments which proved that electricity would not produce the effects upon plants which had been attributed to it; that electrified seeds [6] would not germinate quicker than others. These experiments, reported in the Journal de Physique for December, 1785, were confirmed in the same journal for December, 1786, were given further support in May, 1788, and were finally summarized in "Expériences sur les végétaux." Various other workers later confirmed these researches. It seems to me at present [1800] that the opinion of those who believe that electricity does not favor vegetation is more logical than the contrary opinion.

At the present time (1924), there is still a diversity of opinion concerning the influence of electricity in plant development. The electroculture committee of the British Ministry of Agriculture and Fisheries recommends (1923) the continuation of experiments with high potential discharge,[7] Newman (*38*) in England considers electroculture by the same method as offering practical assurance of increased returns. Baines (*1*) points out a wonderland of electrobiological relationships. On the other hand the experiments of Gerlach and Erlwein (*19, 20*) in Germany and the experiments reported in the first part of this bulletin show no increased growth definitely attributable to electrical treatment.

---

[6] Leighty and Taylor (*31*) report experiments with electrified seed which indicate no advantage gained by treatment.

[7] Typewritten report on file in the Office of Biophysical Investigations, Bureau of Plant Industry.

LITERATURE CITED

(1) BAINES, A. E.
1921. Germination in its electrical aspects. 185 pp., illus. London and New York.

(2) BASTY, F.
1908. Essais d'électroculture tentés à Angers en 1908. (Extrait.) *In* Bul. Soc. Études Sci. Angers, ann. 37 (1907), pp. 87–92.

(3) BERTHELOT, M.
1889. Recherches nouvelles sur la fixation de l'Azote par la terre végétale. Influence de l'électricité. *In* Compt. Rend. Acad. Sci. [Paris], tome 109, pp. 281–287.

(4) BLACKMAN, V. H.
1918–1924. [Electrical treatment of fields.] Gt. Brit. Min. Agr. and Fisheries Interim Rpts. 1–6 (1917–23). [Mimeographed.]

(5) 1924. Field experiments in electro-culture. *In* Jour. Agr. Sci., vol. 14 (1923), pp. 240–267.

(6) ——— and JØRGENSEN, I.
1917. The overhead electric discharge and crop production. *In* Jour. Bd. Agr. [London], vol. 24, pp. 45–49, illus.

(7) ——— and LEGG, A. T.
1924. Pot-culture experiments with an electric discharge. *In* Jour. Agr. Sci., vol. 14, pp. 268–286, illus.

(8) ——— ——— and GREGORY, F. C.
1923. The effect of a direct electric current of very low intensity on the rate of growth of the coleoptile of barley. *In* Proc. Roy. Soc., London, ser. B, vol. 95, pp. 214–228, illus.

(9) BRESLAUER, M.
1912. Amount of energy needed for electro-culture. *In* Electrician, vol. 69, pp. 889–890.

(10) BRIGGS, L. J., and SHANTZ, H. L.
1913. The water requirement of plants. I. U. S. Dept. Agr., Bur. Plant Indus. Bul. 284, 49 pp., illus.

(11) 1914. Relative water requirements of plants. *In* Jour. Agr. Research, vol. 3, pp. 1–63, illus.

(12) CHREE, C.
1910. Atmospheric electricity. *In* Enc. Brit., vol. 6, pp. 860–870, illus.

(13) CLAUSEN.
1911. Die Erfolge der Elektrokultur in Hedewigenkoog. *In* Landw. Wchnbl. Schles.-Holst., Jahrg. 61, pp. 83–86.

(14) DUDGEON, E. C.
[1912]. Growing crops and plants by electricity. 36 pp., illus. London.

(15) ELFVING, F.
1882. Ueber eine Wirkung des galvanischen Stromes auf wachsende Wurzeln. *In* Bot. Ztg., Jahrg. 40 (1881), pp. 257–264, 273–278.

(16) FITCHNER, E.
1861. Agronomische Zeitung, 1861, p. 550. [Not seen. Reference from Bruttini, A., L' influenza dell' elettricità sulla vegetazione, p. 148, Milano, 1912.]

(17) GASSNER, G.
1907. Zur Frage der Elektrokultur. *In* Ber. Deut. Bot. Gesell., Bd. 25, pp. 26–38, illus.

(18) 1909. Pflanzenphysiologische Fragen der Elektrokultur. *In* Mitt. Deut. Landw. Gesell., Jahrg. 24, pp. 5–7.

(19) Gerlach, M., and Erlwein, G.
1910. Versuche ueber Elektrokultur. *In* Elektrochem. Ztschr., Jahrg. 17, pp. 31–36, 66–68, illus.

(20) 1910. Versuche über die Einwirkung der Elektrizität auf das Pflanzenwachstum. *In* Mitt. K. Wilhelms Inst. Landw. Bromberg, Bd. 2, pp. 424–453, illus.

(21) Grandeau, L.
1878. De l'influence de l'électricité atmosphérique sur la nutrition des plantes. (Extrait.) *In* Compt. Rend. Acad. Sci. [Paris], tome 87, pp. 60–62, 265–267, 939–940.

(22) Höstermann.
1910. Geschichte und Bedeutung der Elektrokultur unter Berücksichtigung der neueren Versuche. *In* Arch. Deut. Landw., Jahrg. 34, pp. 535–570.

(23) Holdefleiss.
1885. Elektrische Kulturversuche. *In* Centbl. Agr. Chem., Jahrg. 14, pp. 392–393.

(24) Homberger, E.
1914. Behandlung von Pflanzen mit Hochfrequenzströmen. *In* Umschau, Jahrg. 18, pp. 733–735, illus.

(25) Honcamp, F.
1907. Die Anwendung der Elektrizität in der Pflanzenkultur. *In* Fühling's Landw. Ztg., Jahrg. 56, pp. 490–499.

(26) Jørgensen, I., and Priestley, J. H.
1914. The distribution of the overhead electrical discharge employed in recent agricultural experiments. *In* Jour. Agr. Sci., vol. 6, pp. 337–348, illus.

(27) Kinney, A. S.
1897. Electro-germination. Mass. Hatch Agr. Exp. Sta. Bul. 43, 32 pp., illus.

(28) Kövessi, F.
1912. Influence de l'électricité à courant continu sur le développement des plantes. *In* Compt. Rend. Acad. Sci. [Paris], tome 154, pp. 289–291.

(29) Leicester, J.
1892. The action of electric currents upon the growth of seeds and plants. *In* Chem. News (London), vol. 65, p. 63, illus.

(30) 1892. Action of an electric current upon the growth of seeds. *In* Chem. News (London), vol. 66, p. 199.

(31) Leighty, C. E., and Taylor, J. W.
1924. Electrochemical treatment of seed wheat. U. S. Dept. Agr. Circ. 305, 7 pp., illus.

(32) Lemström, S.
1904. Electricity in agriculture and horticulture. 72 pp., illus. London and New York.

(33) Lodge, O.
1908. Electricity in agriculture. *In* Nature, vol. 78, pp. 331–332, illus.

(34) 1911. The mode of conduction in gases illustrated by the behaviour of electric vacuum valves. *In* Phil. Mag. and Jour. Sci., ser. 6, vol. 22, pp. 1–7, illus.

(35) Maccagno, J.
1880. Influenza dell' elettricità atmosferica sulla vegetazione della vite. *In* Staz. Sper. Agr. Ital., vol. 9, pp. 83–89.

(36) Mascart, E. E.
1876. Traité d' électricité statique. 2 vol., illus. Paris.

(37) Newman, J. E.
1911. Electricity as applied to agriculture. *In* Electrician, vol. 66, pp. 915–916, illus.

(38) 1922. Electricity and plant growth. *In* Standard handbook for electrical engineers, Ed. 5, pp. 1810–1811. New York.

(39) PAULIN.
1892. De l' influence de l'électricité sur la végétation. Montbrison. [Not seen. Reference from Bruttini, A., L'influenza dell'elettricità sulla vegetazione, p. 216. Milano. 1912.]

(40) PLOWMAN, A. B.
1902. Certain relations of plant growth to ionization of the soil. *In* Amer. Jour. Sci., ser. 4, vol. 14, pp. 129–132, illus.

(41) 1903. Electromotive force in plants. *In* Amer. Jour. Sci., ser. 4, vol. 15, pp. 94–104, illus.

(42) PRIESTLEY, J. H.
1907. The effect of electricity upon plants. *In* Proc. Bristol Nat. Soc., ser. 4, vol. 1 (1906), pp. 192–203.

(43) 1910. Overhead electrical discharges and plant growth. *In* Jour. Bd. Agr. [London], vol. 17, pp. 16–28.

(44) ROSS, W.
1844. Galvanic experiments on vegetation. U. S. Comr. Patents Rpt., vol. 27, pp. 370–373, illus.

(45) SENEBIER, J.
[1800.] Physiologie végétale. 5 vols. Genève.

(46) SOLLY, E.
1846. The influence of electricity on vegetation. *In* Jour. Hort. Soc. London, vol. 1 (1845), pp. 81–109.

(47) WILSON, C. T. R.
1923. Atmospheric electricity. *In* Glazebrook, R., A dictionary of applied physics, vol. 3, pp. 84–107, illus.

(48) WOLLNY, E.
1888–1893. Elektrische Kulturversuche. *In* Forsch. Agr. Phys., Bd. 11, pp. 88–112, 1888; 16, pp. 243–267, 1893.

ORGANIZATION OF THE UNITED STATES DEPARTMENT OF AGRICULTURE

December 22, 1925

Secretary of Agriculture	W. M. JARDINE.
Assistant Secretary	R. W. DUNLAP.
Director of Scientific Work	———
Director of Regulatory Work	WALTER G. CAMPBELL.
Director of Extension Work	C. W. WARBURTON.
Director of Information	NELSON ANTRIM CRAWFORD.
Director of Personnel and Business Administration	W. W. STOCKBERGER.
Solicitor	R. W. WILLIAMS.
Weather Bureau	CHARLES F. MARVIN, *Chief.*
Bureau of Agricultural Economics	THOMAS P. COOPER, *Chief.*
Bureau of Animal Industry	JOHN R. MOHLER, *Chief.*
Bureau of Plant Industry	WILLIAM A. TAYLOR, *Chief.*
Forest Service	W. B. GREELEY, *Chief.*
Bureau of Chemistry	C. A. BROWNE, *Chief.*
Bureau of Soils	MILTON WHITNEY, *Chief.*
Bureau of Entomology	L. O. HOWARD, *Chief.*
Bureau of Biological Survey	E. W. NELSON, *Chief.*
Bureau of Public Roads	THOMAS H. MACDONALD, *Chief.*
Bureau of Home Economics	LOUISE STANLEY, *Chief.*
Bureau of Dairying	C. W. LARSON, *Chief.*
Fixed Nitrogen Research Laboratory	F. G. COTTRELL, *Director.*
Office of Experiment Stations	E. W. ALLEN, *Chief.*
Office of Cooperative Extension Work	C. B. SMITH, *Chief.*
Library	CLARIBEL R. BARNETT, *Librarian.*
Federal Horticultural Board	C. L. MARLATT, *Chairman.*
Insecticide and Fungicide Board	J. K. HAYWOOD, *Chairman.*
Packers and Stockyards Administration	JOHN T. CAINE, *in Charge.*
Grain Futures Administration	J. W. T. DUVEL, *in Charge.*

This bulletin is a contribution from

Bureau of Plant Industry	WILLIAM A. TAYLOR, *Chief.*
Office of Biophysical Investigations	G. N. COLLINS, *Senior Botanist in Charge.*

ADDITIONAL COPIES
OF THIS PUBLICATION MAY BE PROCURED FROM
THE SUPERINTENDENT OF DOCUMENTS
GOVERNMENT PRINTING OFFICE
WASHINGTON, D. C.
AT
10 CENTS PER COPY
▽

Spotted something that isn't quite right?

Let us know and we'll fix it for future reprints. And as a thank you for your diligence, we'll send you a free Sothis book of your choice.

publisher@sothis.press

This page intentionally left blank

Made in United States
Orlando, FL
05 November 2026

90231083R00025